ANYTHING BUT ORDINARY

PECULIAR NATURE

BY JOHN MICKLOS, JR.

CAPSTONE PRESS
a capstone imprint

Capstone Captivate is published by Capstone Press,
an imprint of Capstone.
1710 Roe Crest Drive
North Mankato, Minnesota 56003
www.capstonepub.com

Library of Congress Cataloging-in-Publication Data
ISBN: 978-1-4966-8464-6 (library binding)
ISBN: 978-1-4966-8484-4 (eBook PDF)

Photo Credits

Getty Images: AFP, 34, JIJI PRESS, 6, Patrick AVENTURIER, 39, Warren Faidley, 11; iStockphoto: Luca Ladi Bucciolini, 15; Science Source: ROGER HILL, 19, 20; Shutterstock: argus, (sparks) Cover, design element, ArtMari, (tornado) Cover, Asif Islam, top 17, Augusto Cabral, 9, Bambuh, (segull) 37, chamski, 5, chiakto, 43, Christian Delbert, 10, Claud B., 45, Colin Dewar, 14, corbac40, 7, Daniel Prudek, 25, Daryna Kurinna, (mountain) 27, Doug Meek, 44, Egor Shilov, (coral) 37, Eva Mont, (falls) Cover, Extezy, (mountains) 27, Frans Delian, 8, Galyna Andrushko, 5, Graeme Shannon, 24, Guitar photographer, 40, iacomino FRiMAGES, 33, james_stone76, 41, Jason Patrick Ross, 5, Jeff Holcombe, 28, Jim Mallouk, 30, Lemberg Vector studio, (shell) 37, Mihai_Andritoiu, 5, MikeDrago.cz, (ice) Cover, Mumemories, 5, nico99, 29, Peter Hermes Furian, (map) 27, Przemyslaw Skibinski, 5, Rauf Aliyev, (palm tree) 37, Rich Carey, bottom 37, Robert Mutch, bottom 17, Robusta, 1, Shaun Jeffers, 23, Sopotnicki, 38, Studio Barcelona, 31, superjoseph, 5, Timaldo, 26, Tunatura, 36, Vadim Petrakov, 13, Vibrush, (fish) 37, Vlad61, 35

Editorial Credits

Editor: Mandy Robbins; Designer: Kyle Grenz; Media Researcher: Kelly Garvin; Production Specialist: Kathy McColley

Printed and bound in the USA.

PA117

TABLE OF CONTENTS

Words in **bold** are in the glossary.

Wild and Wonderful Nature

Sometimes we take nature for granted. The sun rises each morning. It sets each night. Seasons change throughout the year. We assume that nature will provide enough rain to keep plants, animals, and humans alive. We don't often think about the extremes of nature.

Sometimes we forget the beauty and power of nature. Some natural forces are so powerful they can be deadly. Some things in nature are so odd we can hardly believe they are real.

Are you ready for a peculiar adventure? Be amazed by twin tornadoes and **albino** trees. Learn about remarkable waterfalls and walls of water. Explore one of the world's tallest mountains and deepest canyons. Meet the world's largest living organism. Prepare to explore nature's weirdest sights.

Seven Natural Wonders of the World

Nature offers thousands of amazing sights across the world. These seven have been named the Seven Natural Wonders of the World:

1 **Great Barrier Reef**
2 **Aurora Borealis**
3 **Victoria Falls**
4 **Grand Canyon**
5 **Harbor of Rio de Janeiro**
6 **Mount Everest**
7 **Paricutin**

Walls of Water

Imagine walls of water 100 feet (30.5 meters) high. That's about as tall as a 10-story building! Now imagine those waves racing toward you at speeds of up to 500 miles (805 kilometers) an hour. That's as fast as an airplane travels. This frightening natural wonder is a tsunami.

A tsunami wave crashes into a roadway in Japan on March 18, 2011.

Underwater earthquakes trigger most tsunamis. An underwater earthquake displaces huge amounts of water. This water races across the ocean in a series of waves. The waves grow in height as they draw near land. They crash with enormous force. The waves can destroy buildings and flip over cars. They can cause flooding far inland. The effects may be felt as far as 3,728 miles (6,000 km) away from the site of the earthquake.

How Does a Tsunami Happen?

Tsunami waves grow as they reach the coast.

Tsunamis are among nature's deadliest forces. On December 26, 2004, a massive tsunami rushed across the Indian Ocean. The giant waves affected 12 countries in Asia and Africa. More than 225,000 people died. Millions more lost their homes.

The country of Indonesia suffered the most damage. The tsunami nearly wiped out the city of Banda Aceh on the island of Sumatra. In all, more than 500,000 Indonesians were forced from their homes. This tsunami ranks fifth among the deadliest natural disasters in modern times. It caused an estimated $10 billion in damages.

The December 26, 2004, tsunami left many coastlines in Indonesia with heaps of rubble.

Nuclear Nightmare

A major tsunami struck Japan in 2011. It damaged a **nuclear** power plant in the town of Fukushima. Nuclear **radiation** escaped from the damaged plant. Nearly 50,000 people left their homes because of high radiation levels in the area. Some still haven't been able to return. Experts say it will take 30 to 40 years to complete a full cleanup of the site.

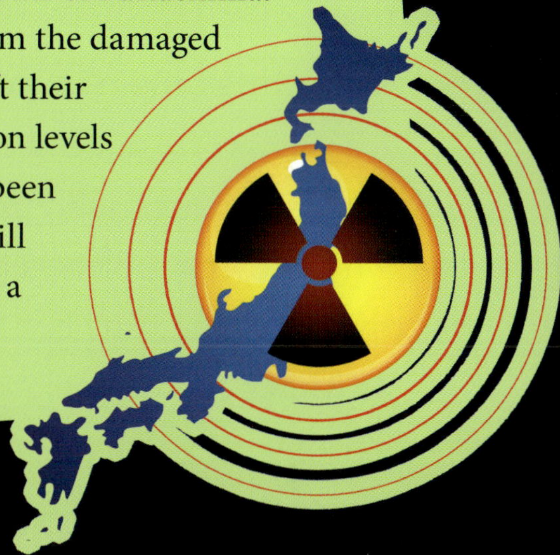

Tsunamis don't happen very often. It is rare to have more than two per year. Today, scientists can use computer models to predict possible tsunamis. These programs use information about earthquake activity in the oceans to predict where a tsunami could hit. This gives people time to prepare. As of now, we have no way to prevent the deadly walls of water from forming or striking land.

Deadly Frozen Droplets

Getting caught in the rain or snow is no fun. It's even worse getting caught in a hailstorm. Hail is a type of solid **precipitation**. It forms when water droplets get swept up to the top of a thundercloud. The low temperatures of the high **altitudes** cause droplets to combine and freeze. Then gravity pulls them down to Earth. The larger the hail, the faster it falls. Large hail pellets can reach speeds of 100 miles (161 km) per hour!

The National Weather Service defines severe hail as measuring 1 inch (2.5 centimeters) or more in diameter. Traveling at high speeds, this hail can damage cars and houses and even kill people.

Hail can seriously damage a car's body.

Hailstones larger than baseballs fell in Joplin, Missouri, on May 22, 2011.

Fact!

In 2010, the largest hailstone on record fell in South Dakota. It measured 8 inches (20 cm) across. That's almost as big as a volleyball.

Some of history's deadliest hailstorms have taken place in India and Bangladesh. One storm in India killed 246 people in 1988. In June 2017, destructive hailstorms struck across the United States. Overall, the storms caused more than $2.5 billion in damages.

Wonderful Waterfalls

Do you love the roaring rush of a mighty waterfall? Have you ever stopped to think about how waterfalls form? They often start when fast-moving water in a river flows from an area of hard rock to softer rock. Over thousands of years, the water wears down the soft rock. The hard rock becomes a ledge from which water falls. After thousands or millions of years, these waterfalls grow.

The world's tallest waterfall is Angel Falls in Venezuela. The water there crashes down 3,212 feet (979 m). That's about the length of 10 football fields. It's also 15 times as tall as Niagara Falls, the most famous waterfall in North America. The water from Angel Falls drops so far and fast that it turns to mist along the way. When it reaches the bottom, it turns back to water.

Angel Falls, Venezuela

Angel Falls is the world's tallest waterfall, but a waterfall in Lebanon may be the oldest. The Baatara Gorge Waterfall in Lebanon dates back to the days of the dinosaurs. Water has been rushing over this waterfall for around 160 million years. Dinosaurs may have stopped to drink at the bottom.

Today, the water of the Baatara Gorge Waterfall drops more than 800 feet (244 m) behind three natural bridges. This makes it one of the world's most stunning waterfalls. People come from all over the world to hike across the natural bridges. Others stand on nearby hills and enjoy the great view of the mighty waterfall. Most visitors come during March or April. Snow from surrounding hills melts at that time of year. It swells the waterfall's flow and creates an even more impressive sight.

Niagara Falls partially frozen in winter

Baatara Gorge Waterfall, Lebanon

Frozen Niagara Falls

About 30 million people visit Niagara Falls each year. This mighty waterfall sits on the border between New York state and Ontario, Canada. The water moves so fast that it rarely freezes, even in the coldest winters. A cold snap in January 2019 froze over much of the falls. Visitors braved bitterly cold temperatures to take photos of sheets of ice where water usually flowed.

Albino Redwoods

Picture a tree as tall as a football field is long! That's how tall the mighty redwoods of California can grow. Some reach heights of 350 feet (107 m). These trees grow slowly over a long period of time. They can live 3,000 years.

Hundreds of thousands of people travel to California each year to see the redwoods. But very few get to see the rare albino redwoods. These trees are also called "ghost" redwoods. Albino redwoods cannot produce **chlorophyll**, so their needles are white instead of green. The roots of these trees connect with neighboring trees. That allows them to get enough food to survive. Because they can't make their own food, they are fairly small trees.

Fact!

In 2018, the Lonely Planet travel team named the California redwood forests the top tourist spots to visit in the United States.

Sequoia and Kings Canyon National Parks in California are home to many giant redwood trees.

Very few albino redwoods exist. Estimates range from just 25 to nearly 400. They are scattered among the other redwoods in the forests. Park workers keep the locations of the albino redwood trees a secret. They want to protect the trees from tourists who might damage them.

Albino redwoods attach their roots to giant redwoods to get nutrients.

Twin Tornadoes

Tornadoes usually start with a thunderstorm. Sometimes the wind's direction and speed change high up in the storm. This causes the air to swirl. Meanwhile, air rises up from the ground. It pushes the swirling mass of air into a vertical funnel. The funnel sucks up more air from the ground and stretches lower and lower.

Once the funnel touches the ground, it is a tornado. On average, more than 1,200 tornadoes occur in the United States each year. The states most often struck by tornadoes are Texas, Kansas, Oklahoma, Florida, and Nebraska.

Tornadoes come as no surprise to people living in Nebraska. About 50 touch down there each year. But imagine seeing two tornadoes at one time! That would be really scary. In June 2014, rare twin tornadoes bore down on the town of Pilger, Nebraska. Pilger is a tiny village of just a few hundred people.

Twin tornadoes roared outside of Pilger, Nebraska, on June 16, 2014.

The Pilger tornadoes developed about 6 miles (10 kilometers) southwest of the town.

Three separate tornadoes formed near Pilger in 2014. One quickly weakened. The remaining two swept through the town. They touched down about 1 mile (1.6 km) apart late in the afternoon. Most tornadoes last for less than 15 minutes. One of these tornadoes skipped along the ground for more than an hour. The tornadoes were ranked at EF4 on the Enhanced Fujita (EF) scale. This means that winds were estimated to be up to 200 miles (322 km) per hour.

One tornado roared right through the center of town. Like many tornadoes, it left behind what seemed like random damage. The storm crumpled some buildings into piles of rubble. It left others untouched. The tornado uprooted trees and crushed cars. It destroyed one of the town's churches. It severely damaged the middle school and the bank. In all, the tornado left two people dead and at least 20 injured.

The Enhanced Fujita Scale

Scientists use the EF Scale to rate the strength of tornadoes. The scale is based on the damage caused by the tornadoes' wind gusts.

EF Rating	Wind Speeds
0	wind gusts up to 85 miles (137 km) per hour
1	wind gusts 85 to 110 miles (137 to 177 km) per hour
2	wind gusts 110 to 135 miles (177 to 217 km) per hour
3	wind gusts 135 to 165 miles (217 to 266 km) per hour
4	wind gusts 165 to 200 miles (266 to 322 km) per hour
5	wind gusts over 200 miles (322 km) per hour

Glowing Caves

Most caves are dark. It's hard to see anything without a flashlight. But the Waitomo Caves in New Zealand glow with a strange light all their own. What causes that glowing light deep beneath Earth's surface? Thousands of glowworms live on the ceilings of the caves. They spin silky glowing threads that attract insects. The insects get trapped in their threads, much like they do in a spider's web.

Every year, people from all over the world visit the Waitomo Caves. Visitors explore the caves by boat, by raft, or on foot. One tour even allows visitors to explore the caves on a zipline. The highlight of the caves is Glowworm Grotto, where the ceiling glows like a starlit sky. The caves also have remarkable rock formations.

Sound travels well in the caves. This makes them good for musical shows. The Cathedral Chamber of the caves hosts underground concerts. Each December, the Waitomo Caves Choir performs there.

Fact!

The light that comes from glowworms is called bioluminescence. That is the same way fireflies glow.

Mighty Mountains

You're walking along a grassy plain in Tanzania, Africa. Giraffes wander by. So do elephants and lions. Everything is flat. The temperature is 80 degrees Fahrenheit (27 degrees Celsius). Then, in the distance, a mighty mountain appears through the mist. There is snow on the peak. You're looking at Mount Kilimanjaro. At 19,341 feet (5,895 m), it is the tallest mountain peak in Africa. It is also the world's tallest mountain that is not part of a mountain range.

Mount Kilimanjaro

the Himalayas

Mountains rarely stand alone. Most are tucked among other mountains in what are called ranges. That includes most of the world's tallest mountains. The very tallest is Mount Everest. It is part of the Himalayas that stretch across parts of central and southern Asia.

Fact!

The Himalayas include more than 110 peaks rising more than 24,000 feet (7,315 m) tall. No other continent has a single mountain reaching that height!

Mount Kilimanjaro formed due to volcanic activity deep beneath Earth's surface. Over millions of years, the mountain was pushed up to its current height. Visitors come from all over the world to see this mighty mountain. They enjoy seeing snow in the heart of Africa. Temperatures at the peak are cold enough to let snow crown the mountain year-round. That has earned Mount Kilimanjaro the nickname "The Shining Mountain."

About 30,000 people climb Mount Kilimanjaro each year.

An Easier Climb

Climbing Mount Kilimanjaro is much easier than climbing mountains in the Himalayas. For one thing, temperatures at Kilimanjaro are warmer. Temperatures at the base of the mountain rarely drop below 45°F (7.2°C). Temperatures in the Himalayas often drop far below zero. Also, Kilimanjaro stands nearly 10,000 feet (3,048 m) shorter than Mount Everest.

The Tallest Mountains on Each Continent

North America

Mount Denali
20,320 feet/6,194 m

Europe

Mount Elbrus
18,510 feet/5,642 m

Asia

Mount Everest
29,035 feet/8,850 m

South America

Mount Aconcagua
22,838 feet/6,961 m

Africa

Mount Kilimanjaro
19,341 feet/5,895 m

Antarctica

Mount Vinson Massif
16,050 feet/4,892 m

Australia

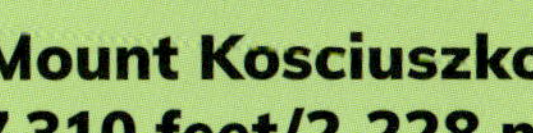

Mount Kosciuszko
7,310 feet/2,228 m

These Plants Eat Meat!

Most plants use the sun's energy to produce the sugar they need as food. This process is called **photosynthesis**. A few types of plants get food in a different way. They eat meat!

These plants have several ways to trap insects. Some have sticky leaves. When an insect touches the leaves, it gets trapped. The plant then slowly digests the insect. Other plants have leaves shaped like pits that collect water at the bottom. Insects slip down the leaves into the water where they drown.

A fly drowns in the liquid inside a pitcher plant.

A Venus flytrap snags a meal.

The most famous type of meat-eating plant is the Venus flytrap. This plant opens its leaves wide. The leaves have short, stiff hairs. These hairs are called triggers. When an insect touches the leaves, the sides of the leaf snap shut. The trap closes tightly around the insect.

Fact!

There are at least 600 types of meat-eating plants.

A Truly Grand Canyon

The Grand Canyon is among the Seven Natural Wonders of the World. It formed over millions of years by the **erosion** of water passing through the rocks. It measures more than 1 mile (1.6 km) deep.

Imagine white-water rafting between cliffs that seem to touch the sky. Adventures such as this draw at least 6 million people to the Grand Canyon National Park each year.

White-water rafters in the Grand Canyon

Visitors ride mules from the bottom of the Grand Canyon to the top.

The Grand Canyon National Park offers something for visitors of all ages. Just looking out over the canyon is spectacular. The views at sunrise and sunset are stunning. Visitors also enjoy many activities in the park. In addition to white-water rafting, visitors can hike or ride horses and mules through the canyon. Some camp on the canyon's rim or at the canyon's base.

The Grand Canyon has at least three **ecosystems**. Each has different plants and animals. The desert ecosystem lies at the base of the canyon. Scorpions, snakes, and sheep live there, as well as many types of fish in the Colorado River. Higher up is the pinyon-juniper woodland ecosystem. That area has mostly pine and juniper trees. The ecosystem nearest the canyon's rim is the boreal and pine forest. This area has aspen, pine, spruce, and fir trees. Deer, squirrels, mountain lions, and other animals roam there.

Protecting Natural Treasures

The National Park Service was created in 1916 to protect the United States' natural treasures. These include the Grand Canyon, the sandstone arches in Utah, and the giant redwood trees of California. Today, the National Park System includes more than 400 sites in all 50 states. In all, these parks cover more than 84 million acres (34 million hectares). Of these, 62 have "National Park" as part of their names.

Many Grand Canyon visitors enjoy seeing the park's waterfalls. The water of the Havasu Creek comes from deep underground. It picks up minerals that reflect sunlight. This makes the water a bright blue-green color. The creek twists and turns through the Havasu Canyon and splashes down a series of waterfalls. The Havasu Falls are the most famous.

Havasu Falls

See It from Outer Space

The Great Barrier **Reef** is the world's biggest structure made up of living things. It can even be seen from space! No wonder it is one of the Seven Natural Wonders of the World. The reef is located in the Coral Sea off the northeast coast of Australia. It stretches for more than 1,429 miles (2,300 km). It's about as big as the entire country of Japan.

From space, the Great Barrier Reef looks like little blue dots off the coast of Australia.

A green turtle swims through the Great Barrier Reef.

The Great Barrier Reef includes about 3,000 different types of coral. Many types of fish and birds call these reefs home. One is the bright orange and white clownfish, the star of the movie *Finding Nemo*. Several at-risk creatures also live near the reefs. These include green turtles. The Great Barrier Reef is the world's largest **breeding ground** for green turtles.

A scuba diver swims through a school of fish near the Great Barrier Reef.

Each year, about 2 million people visit the Great Barrier Reef. They swim among schools of fish. They admire the bright coral through glass-bottomed boats. But humans also threaten the reef's health. Overfishing harms the reef. So does increased traffic from oil tankers.

Climate change and extreme weather also threaten the long-term health of the reef. As much as half of the reef has died since 2016 from hotter than normal ocean temperatures.

Fun Figures About the Great Barrier Reef

400 types of coral

4,000 types of mollusks

240 types of birds

900 islands

1,500 types of fish

Humans can help save the reef. Programs are underway to protect and restore this natural wonder. The Australian government has set up a long-term plan to protect the reef. This plan extends until 2050. In 2018, the government set aside $500 million in extra funding to address threats to the reef. The goal is to make sure the reef survives and thrives.

When reefs die, they lose their bright colors.

Mud Volcanoes

Volcanoes have caused some of the world's worst natural disasters. Mud volcanoes, on the other hand, rarely pose a danger to people. They do not produce **lava**, only mud. Most are quite small. Each year, more than 4 million people travel to Yellowstone National Park in Wyoming to see its mud volcanoes. These volcanoes do not erupt. They simply bubble and steam.

Steam erupts from a mud volcano in Yellowstone National Park.

The mud from the mud volcano eruption in Indonesia has covered entire towns and caused great damage.

Still, sometimes the mud blasts can kill people. The world's largest mud volcano is in Indonesia. It stands 2,300 feet high (701 m). In 2006, it spewed mud that killed 14 people.

More than 10 years after it first erupted, the mud volcano in Indonesia remains active. It has destroyed 15 villages and forced more than 40,000 people from their homes. In all, it has caused nearly $3 billion in damage.

Lighting the Night Sky

The dazzling lights known as the aurora borealis and the aurora australis can be awe inspiring. People north of the equator see the aurora borealis, or northern lights. South of the equator, people see the aurora australis, or southern lights. These lights color the night sky in pink, violet, yellow, blue, and green. This light show is caused from bits of matter that come all the way from the sun. The best places to see the lights are in the far northern and far southern parts of the world.

Iceland has some of the most stunning views of the northern lights.

The southern lights dance in the sky near Betsey Island in southeast Australia.

Each year, thousands of people visit Alaska, northern Canada, Norway, Sweden, Iceland, and Finland to view the northern lights. Early spring and early fall often provide the best viewing times. The sunspots that cause the lights run in 11-year cycles. That means some years have stronger and brighter displays than others. The next peak time for seeing the lights will be 2024.

What Causes the Northern and Southern Lights?

The northern and southern lights start out on the sun. Extreme heat on the sun's surface causes it to boil and bubble. Bits of matter escape from the sunspot areas. These bits of matter are known as solar wind. They race through space and reach Earth about 40 hours later. Then they light up the night sky.

Powerful but Fragile

It is often said that people can't control nature. In most cases, that's true. We can't make it rain. We can't stop tornadoes or hailstorms from forming. We can't stop tsunami waves once they have formed. Nature's power can be scary. But it's less scary when we understand how these things happen.

What we do affects nature. Scientists agree that pollution is causing the earth to warm. This climate change threatens some of the world's natural wonders. Some scientists believe Mount Kilimanjaro may soon lose the snow on its peak. Warmer weather may cause it to melt. Climate change also threatens the Great Barrier Reef. Warmer ocean water has already killed large parts of it. Nature is powerful, but fragile. Natural wonders need our care. Without it, they may not survive.

Air pollution hangs over Santiago, Chile. Air pollution makes it difficult for people with lung problems to breathe.

BUT WAIT... THERE'S MORE

AMAZING ARCHES

Like an artist, nature carved all sorts of wonders in the western United States. In Utah, wind, rain, and streams carved a series of sandstone arches and bridges. These include the Landscape Arch, one of the longest natural arches in the world.

double rainbow

DOUBLE RAINBOW

One of nature's most beautiful sights is a rainbow that appears after a rainstorm. Rainbows are common. Much less common are twin—or double—rainbows. The colors of the second rainbow are much fainter. They also appear in reverse order to the primary rainbow.

WILD WIND

The world's windiest place year-round may be Cape Denison on Antarctica. On a normal day, the wind blows between 39 and 54 miles (63 and 87 km) per hour.

Delicate Arch, Arches National Park, Utah

GLOSSARY

albino (al-BYE-noh)—a living thing with no natural coloring

altitude (AL-ti-tood)—how high a place is above sea level

breeding ground (BREED-ing-GRAUND)—a place where animals go to mate

chlorophyll (KLOR-uh-fil)—the green substance in plants that uses light to make food from carbon dioxide and water

climate change (KLY-muht CHAYNJ)—a significant change in Earth's weather patterns over a period of time

ecosystem (EE-koh-sis-tuhm)—a system of living and nonliving things in an environment

erosion (i-ROH-zhuhn)—the wearing away of land by water or wind

lava (LAH-vuh)—magma that comes out of a volcano

nuclear (NOO-klee-ur)—having to do with the energy created by splitting atoms

photosynthesis (foh-toh-SIN-thuh-siss)—the process by which green plants make their food

precipitation (pri-sip-i-TAY-shuhn)—water that falls from the clouds in the form of rain, hail, or snow

radiation (ray-dee-AY-shuhn)—rays of energy given off by certain elements

reef (REEF)—a strip of rock, coral, or sand near the surface of the ocean

READ MORE

Hoare, Ben. *The Wonders of Nature.* New York: DK Publishing, 2019.

Montgomery, Heather L. *Unsolved Mysteries of Nature.* North Mankato, MN: Capstone Press, a Capstone imprint, 2016.

Squire, Ann O. *Tsunamis.* New York: Children's Press, an imprint of Scholastic Inc., 2016.

INTERNET SITES

30 Freaky Facts About the Weather
www.natgeokids.com/uk/discover/geography/physical-geography/30-freaky-facts-about-weather/

Great Barrier Reef Facts for Kids
factsforkids.co/great-barrier-reef-facts-for-kids/

Weather Wiz Kids
www.weatherwizkids.com

INDEX